Foreward

At the beginning of the Covid pandemic, Melbourne, Australia went into lockdown, I was able to move back to my home town in Ballarat and help support my mother, Violet, who was 91 and living by herself with her dog Bonnie for company.

Having an IT background I could immediately see how technology could enhance my mother's quality of life, as she was also experiencing failing eyesight. I introduced her initially to one Google Home Speaker, you could imagine my mother's surprise when it replied to her question "Hey Google what's the time?" and a voice replied with the answer!

From then I was able to introduce the many other brilliant features that Google Home and associated products offered to enhance my mother's quality of life. After two years my mother, at 93, was still very switched on, although at times she needed my help in reminding her what Google Home could do and helping her in setting up tasks. Overall she embraced Google Home and was very proactive in using its many features.

I'm Hoping this help guide can be of assistance to help the many elderly and sight impaired people, similar to my mother. In memory of my beautiful mother, Violet Ethel Lillian Lobley, born in London in 1928, passed away Melbourne Cup Day 2021 (A horse named Very Elegant, spelt Verry Elleegant, won the Melbourne Cup that day which we thought was very fitting).

Table of Contents

Foreward	02
What is needed	04
Create a Gmail account	05
Download Google Home App to SmartPhone	05
Connect the internet to Google Home Speaker from Google Home App	06
Set up Reminders and Events in Google Calendar	15
Google Routines	18
Set up Music account	20
Google Speaker Phone	22
Voice Controlled Appliances	24
Other Useful Products	25
SmartPhones For Seniors	27

What is Needed

Home Internet

Mobile SmartPhone to download Google Home App

Google Home Speaker

Gmail Account

Many households and organisations have the internet connected or available but may not be using it to its full potential. My mother had the internet connected but only used it for phone and emails.
By introducing my mother to Google Home she was able to make use of its many features at a minimal cost (in my mother's example only the price of a Google Mini Speaker), there was no additional cost to her internet usage.

Google, Android, Google Home Speaker and Gmail are trademarks of Google LLC

Technical Information

Create a Gmail Account

1. Go to https://www.google.com/gmail/about/
2. Select *Create an Account*
3. The Sign Up Form will appear
4. Follow Instructions

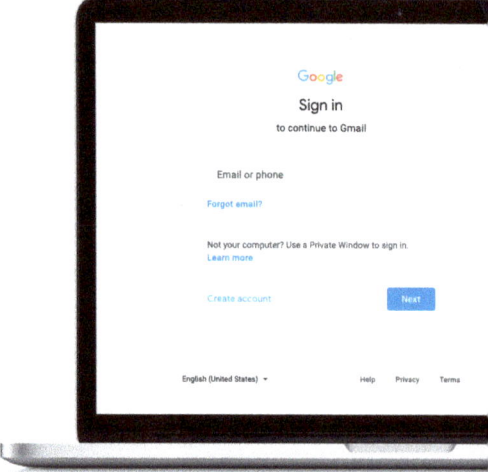

Download Google Home App to a SmartPhone

1. On an available SmartPhone download the Google Home App
2. Google Home Icon appears with Install option
3. Select Install

Note: if you require the Google Home App to be installed onto a PC or Apple Mac there is no direct download available, you will need to investigate if there are any available install options.

For my mother I downloaded the Google Home App and configured it on my iPhone (just needed my smartphone to be connected to my mother's internet when any configuration required).

Google, Google Home and Gome are trademarks of Google LLC

Connect the internet to a Google Speaker from Google Home App

Follow steps in **Google Home App** to connect to available internet and customise **Google Home** settings to User requirements.

Create a User Google Home account, can have multiple speakers set up (ie: livingroom, kitchen, bedrooms etc) in one Google Home App account.

Google offer a range of speakers. For my mother we had the Google Mini Speaker (pictured above right) which provided really good audio at an economical price.

Now that a Google Speaker is connected to the internet and a Google Home account is set up say: "Hey Google…" to request response:

"Hey Google how do you say in Italian: Good morning isn't it a lovely day?"

"Italian translation........"

You can also choose languages other than English to interact with Google.

1. Settings/Google Assistant
2. Manage all Assistant Settings
3. Languages

"Hey Google Heads or Tails?"

"Heads"

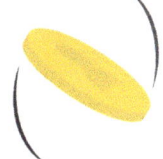

Google can translate languages and you can also choose which language(s) to interact with Google, this can be helpful for grandparents who have migrated from a foreign speaking country.

Google For GrandParents

Google Home can tell jokes and has quizzes available on many subjects, these quizzes can be played by one or multiple players.

Ask Google many questions on interests and facts

Change volume of Google Speaker

Volume of Google Speaker can be adjusted from 1 (low) to 10 (high) verbally, or tap either side of the device to change volume.

Ask Google to repeat answer but more slowly

"Hey Google say that again but more slowly"

Send a message to another speaker

"Hey Google send a message to Bedroom Speaker say: Hey Iris your tea is ready"

Can have multiple speakers on one user account

"OK sending message to bedroom speaker: Hey Iris your tea is ready"

Google For GrandParents | 13

Google can help you remember

"Hey Google remember that my son has my spare set of keys"

"Hey Google where are my spare keys?"

"Your son has your spare set of keys"

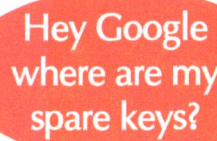

Ask Google to Create a Shopping List

"Hey Google what's on my Shopping List?"

"Hey Google add Milk to my Shopping List"

Set Up Alarms

You can set up a daily or one off alarm either by voice or configured in Google Home App.

1. Hey Google set an alarm for 7 am every day
2. Hey Google set an alarm for 5 am next Thursday

Customise Google Home Alarm
Wake up to favourite Music/Rado or character voice, daily news, perform tasks

Turn on Snooze

"Hey Google Snooze" (Default 10 minutes)

"Hey Google Set Snooze for ... minutes"

Turn off Alarm:
"Hey Google Stop"

Google For GrandParents

Set Up Reminders and Events in Google Calendar

Download Google Calendar onto SmartPhone or other device (can have multiple copies for family members of User's Gmail account).

1. On SmartPhone or other device go to App Install
2. Select Google Calendar
3. Select Install
4. On iPhone Select Settings/Calendar/Account/ Add Account
5. Follow Instructions

Note: Spoken Reminders and Events entered into Google Home by voice will also appear in Google Calendar App (example following pages).

Also, this feature enables any family member or friends to view or enter Reminders and Events from their device if they are logged into that person's account.

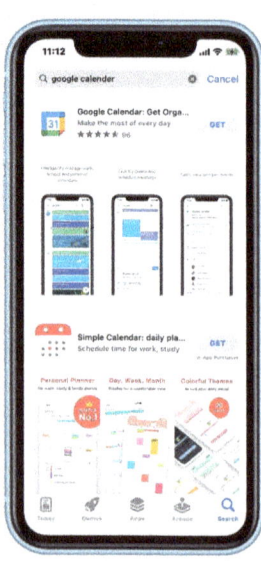

Google and Google Calendar are trademarks of Google LLC

Google For GrandParents | 15

> Hey Google set me a Reminder for 2 pm Friday called Doctor Appointment

> Hey Google set me up an Event to repeat every March 28 called Syd's Birthday

> Hey Google do I have any Reminders for Friday?

> Hey Google have I any Events for the month of March?

Example below of Google Calendar display can be viewed on multiple smartphones so family members and friends can monitor elderly person's activities, and can also enter Reminders and Events into a person's calendar from their own device.

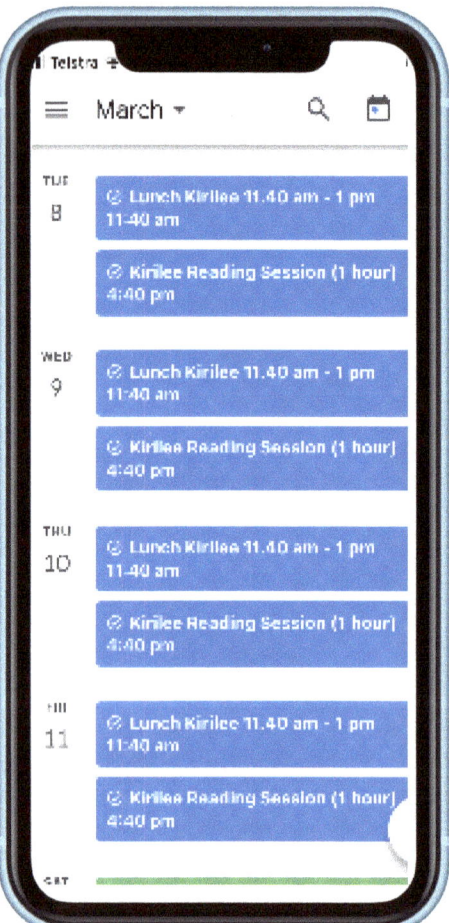

Hi Grandma I'm entering a Reminder for you into your Calendar

Family members and friends can update and view User's Calendar

Hi Grandma today is Wednesday I see you have a book reading at 4.40 pm would you like me to come with you?

Google For GrandParents | 17

Google Routines

From the Google Home App you can also enable Google to announce a Routine to execute with Google announcing a reminder/event by voice on a particular day and time.

Create a daily Routine at 8 am each morning to say: "Hey Violet it's 8 am, time to take your medication". This can be done by setting up the required routine in the Routines section of the Google Home App.

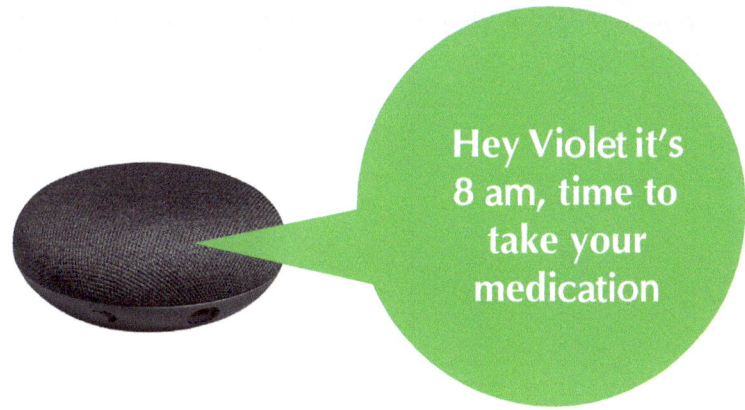

Technical Information

1. Open Google Home App (SmartPhone)
2. Under *Settings select   Google Assistant*
3. *Select   Manage All Assistant Settings*
4. Scroll down and select *Routines*
5. Select *+ New*   top right hand corner
6. Select *+ Add Starter*
7. *Select at Specific Time*

8. Time of Day?
9. Repeat every?
10. Select *Add Starter*
11. Select *+ Add Action*
12. Select *Communicate and Announce*
13. Tick box *Say Something*
14. Enter words you would like Google to say ie: "Hello Violet it's 8 am, time to take your medication"

15. Select Blue Tick in right hand top corner
16. Select *Done*
17. Enter Name of Routine at top of screen
18. Select Blue Tick in right hand top corner
19. Select *Save*

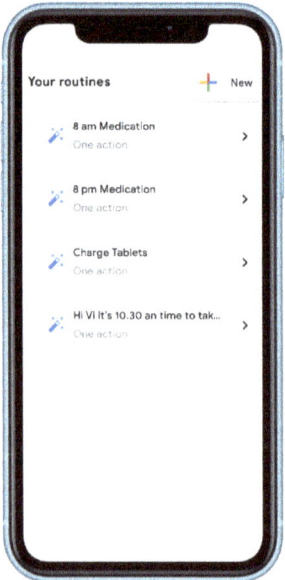

8 am every day Google Announces:

Hello Violet it's 8 am, time to take your medication

Set Up Music Account

My mother enjoyed listening to many songs and programs using a song provider, in her case she used Spotify Premium, but there are other music service providers available. I was able to set up playlists of mum's favourite songs, British comedy shows and books.

I created playlists of songs such as her favourites from The Beatles, Vera Lynn, Harry Secombe, Elton John and British comedy shows such as Dad's Army. For continuity, I named many playlists with my mother's name (name playlists as desired).

Technical Information

1. On SmartPhone or other device install a Music App from the App Install

2. Set up Playlists (if availabe)

3. Choose Music Provider in Google Home Application: *Home/Settings/Music*

Music Servics available on Google Home:
You Tube Music; Spotify; Spotify Premium;
Apple Music; Deezer; iHeart Radio
For my mother we created her Playlists using Spotify Premium (this required a monthly subscription fee) but there are FREE Music options available.

"Hey Google play the playlist Violet's Favourites"

"OK playing playlist Violet's Favourites on Spotify"

"Hey Google play playlist Violet's Favourites Number 2"
"Hey Google play playlist Violet's Favourites Number 3"
"Hey Google play playlist Dad's Army Number 1"
"Hey Google play playlist Violet's Violins"
"Hey Google play playlist Violet's Books"

Turn off Music say: **"Hey Google Stop"**

At times Mum had trouble remembering what playlists to play, we overcame this remotely by me telling mum to place the phone next to the Google Speaker and I would issue commands over the phone.

Play Digital Radio

Hey Google play 3AW radio Melbourne, Australia

Digital Radio from around the world

"Hey Google play ... radio"
"Hey Google what is the news today?"

Turn off Radio say:
"Hey Google Stop"

Google Speaker Phone

Google Home Speaker can make and receive phone calls

Google Home Speaker can be used to make and receive phone calls. Users in Australia must have a Telstra mobile phone acount. This can be set up in the Google Home App under Settings/Communication.

Technical Information

1. Open Google Home App
2. Select *Settings*
3. Scroll down and choose *Communication*
4. Choose Call Provider
(Your Call Provider needs to have this feature available) In Australia you must have a Telstra mobile phone account.
5. Follow instructions

"Hey Google call Janine"

On setup you can sync the Google Home App to the User's own contacts

"Hey Google what is the number for Clip Hairdressing?"

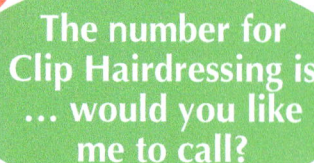

"The number for Clip Hairdressing is … would you like me to call?"

"Yes please"

"Hello this is Clip Hairdressing …"

My mother also chose to have a landline phone with an answering machine

1. The Google Home App can also allow users to answer phone calls hands-free through voice commands. If you have *Inbound Callig* enabled, you can accept calls by saying: **"Hey Google"** then **"Answer the Call"** or tap on the device's touch surface to accept the call.

2. Alternatively, you can choose to say: **"Hey Google Decline the Call"**.

3. End a Call say: **"Hey Google Hangup"**.

Voice controlled Appliances

There are many devices that can be voice operated some of which I set up for mum.
I was able to introduce some voice controlled appliances to help make every day tasks easier as my mother's eyesight was failing.

Air Conditioner
Mitsubishi Split System Air Conditioner (with voice control) has an app that can be downloaded and controlled remotely from a SmartPhone.

Mum could issue voice commands:
"Hey Google turn on thermostat to cool".
"Hey Google set thermostat temperature to 18 degrees celsius" etc.
Family members can also have the app downloaded to their device and can operate settings remotely.

Hey Google turn on air conditioner

Lights
Smart Lights (LIFX) can be installed in any room and operated by voice.

Hey Google turn on Kitchen Light

Smart Television
With Remote Access capability
By using a Smart TV with remote access family members can control the TV remotely and find and play programs that some elderly people would have difficulty in playing. Larger screens benefit sight impaired people and can also be used for Zoom/Skype hookups etc.

Hey Google turn on TV

SMART TV

Other Useful Products
Computer/Laptop/Tablet
(with remote access ability initiated)

Again like a Smart TV with remote access family members can gain remote access to User's tablet/laptops from their device to initiate such things as audio books etc.

Google For GrandParents | 25

Google Activity

https://myactivity.google.com

Google Activity is a tool you can use to monitor data and conversation requests. This was a good tool to see how mum was interacting with Google and I was able to help her when struggling with her commands.

SmartPhones for Seniors

Tips for the elderly and sight impaired to embrace the features of a SmartPhone

My mother obtained an iPhone SE SmartPhone through her government support package. While she was reluctant at first to get a SmartPhone, I convinced her of the many brilliant features a SmartPhone has, especially to assist the elderly and sight impaired.

To mum's delight we were able to configure her iPhone to the extent that all she had to do was ensure it was sufficiently charged each day. Every other action was achieved by her voice commands using Siri, the iPhone voice automated assistant.

1. Turn Siri on (iPhone)

1. Go to Settings on iPhone

2. Scroll down and choose Siri & Search

3. Choose whether to enable Siri with "Hey Siri" or choose to press the Home Button

Hello, I'm Siri your voice activated assistant how can I help you today?

I found saying "Hey Siri" was the best option for my mother to engage with Siri as her eyesight was failing.

2. SmartPhone automatically answers all calls

By turning on Auto Answer Calls, family members and friends were guaranteed to get their calls through to mum, there was no need for mum to fumble for her iPhone or press any buttons.
Warning: this feature will also auto answer unwanted callers (such as scammers), so may not suit everyone.

1. Open Settings on iPhone

2. Scroll down and select Accessibility

3. Scroll down and select Touch

4. Select Call Audio Routing

5. Tap Auto Answer Calls to on

Siri can announce name of caller

1. Go to Settimgs
2. Scroll down: Siri & Search
3. Select Announce Calls
4. Tick Always

6. You can also choose the duration of time the iPhone will ring before it automatically answers the call by tapping plus (+) or minus (-) to the right of seconds (we had mum's timer set to 12 seconds).

Allow calls only from Contacts

1. Go to Settings/Focus/ Do Not Disturb
2. Select Do Not Disturb
3. Allow Notifications
4. Select People
5. Select Calls From
6. Tick All Contacts

WARNING:
This feature will also auto answer unwanted callers (such as scammers), so may not suit everyone.
Don't turn on Auto Answer Calls if this could be a problem!

3. Use Siri to make Calls

Can call any name from phone contact list

Ask Siri to find the phone number of a business and call

Hangup Phone Call: later versions of the iPhone have the ability for Siri to end phone calls.

1. Open Settings on iPhone
2. Select Siri & Search
3. Select Siri Call Hangup
4. Set toggle to on

Well Dear I have to go now: Hey Siri end call

4. SmartPhone Speaker auto on when making and receiving phone calls

1. Open Settings on iPhone
2. Scroll to Accessibility/Touch
3. Scroll to Call Audio Routing
4. Select Speaker

To help with Mum's hearing we had the Phone Speaker on for all calls being received and when calling from her iPhone.

Google For GrandParents | 29

5. Ask if you have any Text Messages?

You can also ask Siri: "Hey Siri do I have any voice messages?"

6. Send a Text Message

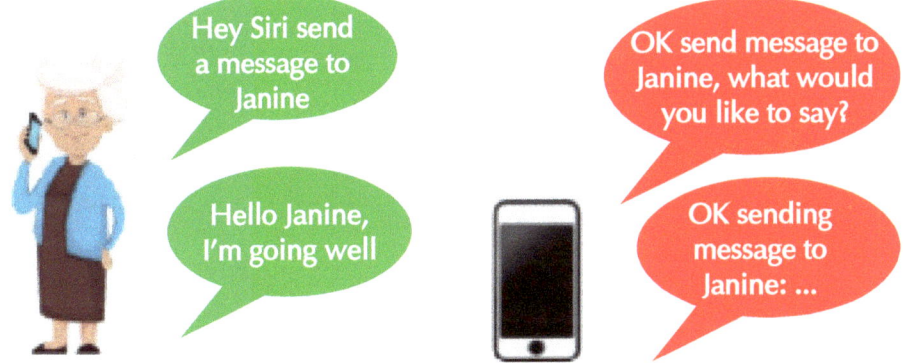

7. Check battey level

8. Divert calls when Phone is busy

Divert call to a family member's phone or answering machine if person on an existing call.

1. Select iPhone Phone Keyboard

2. Enter *67* and phone number you are diverting to
ie: *67*123456789

3. Press Dial and wait for confirmation

Turn off Call Forwarding:
1. Select iPhone Phone Keyboa
2. Enter #61#
3. Press Dial and wait for confirmation

9. Install Google Calendar (see details pages 5 & 15)

Siri has the ability to announce and update Calendar Reminders and Event

Copyright 2022 Stephen Lobley

"All rights reserved. No part of this publication may be reproduced, distributed or transmitted in any form or any means, including photocopying, recording, or other electronic or mechanical methods, without the prio written permission of the publisher, except in the case of brief quotations embodied in critical reviews and cer other noncommercial uses permitted by copyright law.

While all attempts have been made to verify the information provided in this publication, neither the author n the publisher assumes any responsibility for errors, omissions, or contrary interpretations of the subject matte herein.

Adherence to all applicable laws and regulations, including international, federal, state, and local governing professional licensing, business practices, advertising, and all other aspects of doing business in Australia, Ne Zealand, England, the US, Canada, or any other jurisdiction is the sole responsibility of the purchaser or read

Any perceived slight of any individual or organization is purely unintentional.

iPhone and Siri are trademarks of Apple Inc and this book is not endorsed by or affiliated with Apple in any w

Google and Google Docs are trademarks of Google LLC and this book is not endorsed by or affiliated with Google in any way."

Paperback ISBN: 978-0-646-86074-9
Digital ISBN: 978-0-646-86077-0

www.ingramcontent.com/pod-product-compliance
Lightning Source LLC
Chambersburg PA
CBHW040743020526
44107CB00084B/2871